Einstein Anderson
science geek

Lightning
Never Lies

and Other Cases

Seymour Simon
Illustrations by Kevin O'Malley

Volume **2** in the Einstein Anderson Series

SEYMOUR SCIENCE

Published by Seymour Science LLC.

These stories, which have been substantially updated and expanded for a new audience, are based on the Einstein Anderson book originally published in 1980 by Viking Penguin, New York, under the title *Einstein Anderson Shocks his Friends,* and republished in 1997 by Morrow Junior Books, New York, under the title *The Halloween Horror and Other Cases.*

Contact: Seymour Science LLC,
15 Cutter Mill Road, Suite 242,
Great Neck, NY 11021.
www.SeymourSimon.com

www.StarWalkKids.com

ISBN: 978-1-936503-10-0

Contents

Contents

The
Flying Saucer

It was the middle of a hot summer and the town of Sparta should have been quiet and sleepy. Instead, it was buzzing with excitement. It seemed that everywhere you went people were talking about "the news." Morris Janus, a lifelong resident of the town, was going around telling anyone who would listen that he had been kidnapped by aliens—and some people even believed him.

"It's because he was on television," Paloma Fuentes told Einstein Anderson. "On that show, Inside the Action."

Einstein Anderson nodded quietly. He was often quiet when he was thinking and he thought a lot. That's how he got the nickname Einstein—after Albert Einstein, a brilliant thinker and the most famous scientist of the twentieth century.

Ever since he was a little kid, Einstein, whose real name was Adam, had been interested in science. Not only did he know an awful lot, but he was also really good at using science to solve all sorts of puzzles and mysteries. He didn't look much like a science wizard. He was just an average-size twelve-year-old boy with light brown eyes that were a little nearsighted, and his glasses seemed a bit too big for his face.

Now he was sitting on the grass in Brookdale Park, near the center of Sparta, with his friend Paloma. Soccer camp was over for the day and he and Paloma were birding, watching a great blue heron wading in the lake. Both kids were dressed in jeans, T-shirts, and sneakers. Paloma's backpack, which she carried everywhere, lay open on the ground. Out of it she'd pulled a pair of binoculars, a digital camera, and a birding book. She also had a birding app open on her phone—one that played recordings of bird calls.

Paloma had been Einstein's best friend for a couple of years, ever since she had moved to town. She was a little taller than Einstein and had long black hair that she always wore in a ponytail, just like she always wore red canvas high-top sneakers and jeans. Paloma was the only person Einstein knew who got as excited about science as he did, although he thought that sometimes she got too excited. This looked like it might be one of those times.

"Let's prove what a fake he is," Paloma said excitedly, handing the binoculars to Einstein. "I bet we can do it, no problem. Everyone knows there's no such thing as aliens."

"Everyone doesn't know that," Einstein corrected her. "The universe is so vast, there probably is life on other planets. The question is, are they visiting us?"

"Come on, Einstein," Paloma said. She liked to tease him about his nickname. "You don't believe that, do you?"

"I don't know," Einstein replied, holding the binoculars up to his glasses. "We might think it's highly unlikely, but we can't be one hundred percent sure that aliens don't visit the earth."

Just then Einstein's phone rang with the

theme song from Star Wars. The sound startled the heron, which flapped its wings and slowly flew away.

"Great!" Paloma grumbled.

"Sorry," Einstein said, fumbling for the phone in the pocket of his jeans. "I should have turned that off. At least we know herons have good hearing."

"The sound waves in the air probably bounced off the water," Paloma shook her head. "Water reflects sound almost as well as something hard like wood. That's why the heron heard your phone. And I was just going to take a photo."

Einstein nodded as he tapped the screen of his phone to answer it. "Hi Mom," he said. "Uh-huh. Uh-huh. Okay. Yeah, sure." Then he touched the small screen and hung up.

"What?" Paloma asked impatiently. She

could tell from the expression on Einstein's face that he had a surprise in store. "What's up?"

"My mom is doing a story about that Janus guy," he told her. Einstein's mom, Emily Anderson, was a reporter and editor for the Sparta Tribune, the local newspaper. "She's going to interview him this afternoon and she asked if we wanted to go along. She wants us to be there to check out his story."

"Are you kidding?" Paloma cried as she jumped up and started stuffing her gear into the backpack. "What are we waiting for?"

She took out her phone and tapped out a quick text message to her mom.

"Okay," Einstein said as he stood up. "But since we're talking about outer space, first answer this—what do planets like to read?"

Paloma groaned. Among other things, Einstein was famous for his corny jokes.

"I don't know, Einstein," she told him with a frown. He was already laughing.

"Comet books!" he said with a smirk.

They jumped on their bikes and raced to Einstein's house. Emily Anderson was waiting and they all got in the Anderson's car and drove downtown. Mr. Janus was waiting for them at the Tribune offices.

Morris Janus turned out to be a long, thin man with sharp features. His eyes were black and piercing. Einstein's mom led them all into a conference room where they took seats around a large table—Emily Anderson and Morris Janus at one end and Einstein and Paloma at the other.

"I hope you don't mind if my son and his friend sit in," Mrs. Anderson said with a smile. "They were so eager to meet you."

"Oh, I don't mind," Janus answered, giving the two young people a broad smile. "I'm getting used to being a celebrity."

"Getting used to being a fake, you mean," Paloma muttered under her breath. Einstein nudged her under the table.

"Now I know you've been interviewed many times," Emily Anderson said to Janus, rather sweetly. "But if you don't mind starting over from the beginning, I think our readers will really enjoy getting all the details."

"I don't mind at all," Janus said with a broad smile. "Just make sure you spell my name correctly. That's M-O-R-R-I-S. J-A-N-U-S."

"Got it," Emily Anderson told him. Then she switched on a small voice recorder she always carried. "Now, please tell us exactly what happened."

Then Janus repeated, almost word for word,

the story he had told many times before. He said that one night a large, glowing, green spacecraft landed in his backyard. A hatch on the side of the ship opened to reveal a group of alien beings inside. He said that they were about three feet tall with the general shape of humans. They each had a head, two arms, and two legs. Their skin was a faint greenish color that seemed to glow. Their eyes were very round with no pupils, and their ears were pointed.

"They ordered me to get in their flying saucer," he said, almost acting out the scene.

"But how did you understand them?" Emily Anderson asked.

"The saucer people were able to talk to me in English with some kind of translation machine they carried," he replied, nodding earnestly. "In just a couple of seconds we had zoomed off into outer space. They took me to a much bigger ship. That part is kind of foggy. All I know for sure is that two days later I woke up in my backyard in the middle of the night."

"You don't remember anything else?"Mrs. Anderson asked, sounding concerned.

"Just bits and pieces," Janus replied, shaking his head. "I think they gave me some kind of drug so they could look into my brain and find out the secrets of the human race."

"The secret is how did they ever find a brain in that numbskull," Paloma whispered to Einstein. He nudged her under the table again. But Janus didn't hear her.

"Well, I do remember they took me for a walk outside their ship," he said very dramatically. "They put me in a space suit and we went out in the airless vacuum of space," he continued. "We had to be tethered to the ship because we were weightless and might float away. The aliens showed me a platform floating nearby where they were building something. The noise was terrible with all those machines going. I could hear the racket through the helmet of my space suit."

"And what were they building?" asked Mrs. Anderson.

"They told me it was a terrible weapon—a ray gun," Janus answered, looking very grave. "They asked me to warn the people of the world that we had better become peaceful or they would destroy us. Only if we learn the ways of peace will we be invited to join the Galaxy Federation of Intelligent Beings."

Emily Anderson asked questions for a few more minutes, until Mr. Janus ended his story with his return to Earth.

"And so that's my mission," he said very dramatically. "To bring the message of the aliens to the people of Earth. And I'd like to thank you for helping me spread the word."

"Well, that's very interesting," Mrs. Anderson said with a big smile. "Very convincing." She turned to Einstein and Paloma at the far end of the table.

"Kids," she asked. "Do you have anything to ask Mr. Janus?"

Paloma opened her mouth to say something insulting, but Einstein beat her to it.

"Yes, I have one," he said. "My question is— why did you make up this story?"

"Make it up!" Janus cried. "How dare you, young man! My story is true—every word of it."

"Not every word," Einstein replied coolly. "You made one big scientific mistake that proves you were not in outer space."

Can you solve the mystery? What was the scientific error in the story Mr. Janus told?

"I did?" Janus said, sounding suddenly less sure of himself. Then he quickly added, "No, I didn't!"

"Sure you did!" Paloma chimed in, sounding confident. Then she added, "At least I think you did. Tell him, Einstein."

Mr. Janus turned to Emily Anderson. "What kind of newspaper is this?" he asked angrily. "You let children ask the questions? Make wild charges? This interview is over!"

Mrs. Anderson just turned calmly to Einstein. "Adam," she asked. "What mistake did Mr. Janus make?"

"He said that he could hear the racket made by the aliens as they built their space weapon.

He said it was so loud he could hear it through his space-suit helmet."

"So? Have you ever heard aliens build a space weapon?" Janus replied. "It was very, very loud."

"But it couldn't be," Einstein said coolly. "Because you were in the vacuum of outer space—where there is no air."

"So what?" Janus asked, but he sounded nervous.

"Tell him, Paloma," Einstein said.

"So—if there's no air," Paloma cried, "there can't be any sound! Sound waves are carried through air or water or the ground. You said they were building the weapon on a platform floating in space. There couldn't be any sound that reached you."

For the first time, Mr. Janus was speechless. "I . . . uh . . . I . . . I don't care about science!"

he shouted and stormed out of the room.

"Thanks, Einstein," Mrs. Anderson said as she turned off the tape recorder. "I think I have enough to write the story now."

"Too bad Mr. Janus left in such a hurry," Einstein said with a small smile. "There was one more question I wanted to ask him."

"Oh, no!" Paloma groaned. "What now?"

"Why did the farmer take the alien to see a rake?" Einstein asked, already laughing.

"I don't know," Paloma grumbled.

"Because the alien said, 'Take me to your weeder!'"

From: Einstein Anderson
To: Science Geeks
**Experiment: Make up
  a Space Language**

Hi Science Geeks. Here's a challenge that would definitely have stumped that two-faced Mr. Carl Janus who claimed to have been abducted by aliens. Give it a try!

 No one on Earth has ever met someone from another planet, but the Universe is so vast we can't say for sure that Earth is the only planet with intelligent life. In fact, we have been sending television signals into space for many years, with the hope that life forms on other planets would be able to pick them up. But there's one problem: Suppose a civilization on a distant planet could pick up television signals from the earth, how would we communicate with them? We'd need a picture language that can be understood by anyone, no matter what language they speak and whether or not they write. Can you make up a picture language that could be understood by an alien intelligence? A Universal Language?

Here's what you need:

- A pencil
- Some paper
- A friend to help you

Try drawing a series of shapes that could be recognized as symbols for numbers. Then think of other shapes that would stand for addition, subtraction, equals, and so on. Try out your idea by showing your series of shapes to a friend. Tell him or her to pretend that it is a message in code from another civilization and he has to decipher it.

Here's an example:

Philip Morrison, a physicist from Cornell University, has devised a visual number system. Here is how he might send the math problem 1 + 4=5.

Can you decipher these symbols? Can you write another problem using the same picture language?
Show your problem to a friend and see if they can
decode it. Then see if you can make up a picture
code all your own.

You can send a message in code to other kids who read
the Einstein Anderson books at

www.seymoursimon.com/EAcode

Visit there yourself and try to decode other kids'
messages.

The Science Solution

Astronomers are sending messages into space today with the hope that intelligent beings on another planet will pick up the signals and figure out how to decode our messages. They are also scanning the sky with radio telescopes to find out if anyone has received our messages and is sending out a reply. How could we recognize that a signal was sent by an alien civilization? Astronomers have picked up radio signals from outside Earth's atmosphere, but so far they think they have come from natural sources. They hope to recognize a radio signal from an intelligent being by its regularity or by some other feature such as a pattern.

Another difficulty in communicating with a planet or a distant star is that radio transmissions travel at the speed of light, which is 186,000 miles per second. On Earth that is very fast. But at that speed radio waves still take years to reach the stars. Even the closest stars are trillions upon trillions of miles away. You could say "hello" today – the message would take years to get there and then the same number of years for an answer to get back to Earth! Talk about a long conversation. Just to say "hello" back and forth might take decades!

Still, it's worth trying, don't you think? Wouldn't you like to be the first person to chat with an intelligent being from another planet?

A Heavy Problem

It was a Saturday morning at the end of July and a real scorcher, the hottest day of the summer so far. Einstein had gotten out of bed early and packed his swimsuit and towel in his backpack, along with a copy of Popular Science magazine. His family was going up to Carter Lake for the day and he couldn't wait.

After a quick breakfast the whole family piled into the pickup truck. Einstein's dad, Matt Anderson, was driving. His mom was in the passenger seat up front and his little brother Dennis was next to him in the back. Dennis had short brown hair and a face full of freckles. Einstein was texting Paloma, finding out who was already at the lake.

"Do you think the truck would float?" his little brother asked in a loud voice as they drove along the highway.

"That's an interesting question, Dennis," Matt Anderson said from behind the steering wheel. "Maybe we should drive into the lake and find out."

"I don't think so!" Emily Anderson said with a laugh.

Dennis was eight years old and often asked strange questions. As usual, Einstein thought it was a good idea to get his brother interested in science. He pushed his glasses back on his nose and gave it a try.

"If the truck cabin was airtight it might float," he told his brother. "It all depends on the specific gravity."

"Gravity?" Dennis shook his head. "You got it all wrong, Einstein. Gravity makes stuff sink!"

"Not that kind of gravity," Einstein said, trying to be patient.

"There's some other kind of gravity?" Dennis replied. Sometimes Einstein thought Dennis misunderstood him on purpose. "Is there a kind that makes stuff go up? Like anti-gravity in the movies?"

"No, no, no," Einstein said. "Specific gravity is a way to measure how much an object weighs compared to the same amount of water. You take a volume of something—like a cubic centimeter of wood. If it weighs less than a cubic centimeter of water, it will float."

"Cars aren't made out of wood," Dennis laughed. "Everyone knows that!"

"I didn't say cars are made out of wood!" Einstein replied, trying to keep his cool. "I'm saying something will float if it weighs less than the same volume of water."

"Volume?" Dennis said. "That's a good idea. Hey, Mom, can you turn up the volume on the radio?"

Einstein just shrugged. He figured sooner or later some science would get inside Dennis's brain.

Soon they were at the lake. It seemed like the whole town was there, trying to cool off. The Andersons went straight to the bathrooms and changed into their swimsuits then headed for the beach. Paloma was already there with her Aunt Camilla, who lived just up the hill.

"Hey, Einstein!" she called when he showed up, teasing him again about his nickname.

"Hi, Paloma," Einstein replied. "How's the water?"

"It's great. Even Stanley is having fun for once and not making trouble."

"Oh, is he here, too?" Einstein asked. Stanley Roberts was a kid in their grade who was always up to one scheme or another. He thought he was going to be the next Steve Jobs or Mark Zuckerberg—someone who made a billion dollars from new inventions. The problem was Stanley's inventions never seemed to work. Stanley thought that Einstein was his biggest rival, and was always trying to prove that he, Stanley, was the real science genius.

"Yeah, he's around here someplace," Paloma

answered. "Let's go in! I'll race you to the raft!"

They turned and headed for the water where a crowd of kids were splashing and playing the game, "Marco Polo." But they didn't even have a chance to get their toes wet when someone called out Einstein's name.

"Hey, Einstein, wait up!"

It was Pat Hong. Pat was in the same grade as Einstein and Paloma, but they hadn't really been friends until this summer. Pat was the best athlete in school. The tall, muscular kid was a natural at sports, including soccer, basketball, and baseball. Now he came running up to them, in a pair of brown board shorts. He was soaking wet from his toes to his short black hair.

"Hi, Einstein," Pat said again. "I'm glad you're here." Then he added awkwardly, "Oh . . . Hi, Paloma." Pat always got a little shy when he was around Paloma, but she didn't seem to notice.

"Hi, Pat," she answered, "You want to race to the raft?"

"Uh, no," Pat said and seemed even more flustered. "I mean, first I want to tell Einstein . . . I mean, tell you and Einstein . . . I mean . . ." he stopped, then quickly went on, "Stanley's been telling everyone that he's going to get even with you for spoiling all his schemes, like his plan to sell friction-free Rollerblades."

"Or the way he tried to trick you into buying his phony dog vitamins," Einstein added.

"That's right," Pat nodded. "So he says he's going to challenge you to a contest."

"A contest?" Paloma asked. "What kind of contest?"

"A contest to prove who knows more about science!" said a voice nearby.

They all turned and there was Stanley himself. Stanley was a tall, thin kid with short blonde hair and a narrow face. He was wearing a pair of plaid swim trunks and a white T-shirt with his own photo on it. Now he strode over to the group of kids, with an air of importance.

"I've had enough of you going around calling yourself Einstein and acting like you know more than everyone else," he said, standing almost eye-to-eye with Einstein.

"He doesn't call himself Einstein," Paloma objected. "Everyone else does."

"Plus, he does kind of know more than everyone else," Pat said in a low voice.

"No, he doesn't!" Stanley shouted angrily. "And I think we should settle this once and for all."

A group of kids had gathered around them, standing at the edge of the water, eager to see what would happen. Einstein's little brother Dennis ran up and pushed his way through the crowd.

"What's happening?" he asked of no one in particular. "Is there going to be a fight?"

Einstein hadn't said anything at all. But now he spoke.

"No, there isn't going to be a fight," he told Dennis. "And I don't think I know more than everyone else. And I don't think we need a contest. Science is for solving problems, not for showing off."

"Hah!" Stanley cried, turning to the other kids who had gathered around. "The great Einstein Anderson is a coward! Afraid to get shown up as a big fake!"

"Don't let him say that, Einstein," Dennis pleaded.

"Yeah, Einstein," Paloma agreed. "Make him

eat his words."

Einstein looked around, then after a few seconds he nodded. "Okay, we'll have a contest," he said. "But since you're the challenger I get to decide what it is."

"That's fair," Pat said, and a few of the kids nodded.

Stanley looked pretty unhappy. It was clear he hadn't thought of that.

"Okay," he grunted. "I guess so. But no tricks!"

"No tricks," Einstein agreed. "Just science. That's what you wanted, right? So here's the challenge—to see who can lift a heavy weight highest, just by using his knowledge of science. That means no machines—no levers, ramps, or pulleys. Agreed?"

Stanley looked like he was thinking really hard. "Well," he said slowly. "Okay. But what's the weight?"

"Pat," Einstein answered with a smile.

"Me?" Pat cried. "You're kidding!"

"No," Einstein told him. "And I'll make it easier for you, Stanley. I won't lift Pat. I'll let Dennis do it for me."

"Me?" Dennis said excitedly. "Okay!"

Stanley grinned wickedly.

"Okay, Einstein," he said with a sneer. "You're on. I think I can lift Pat higher than your baby brother can. Watch this!"

And with that, Stanley stepped over to Pat, put his arms around the big kid, and heaved with all his might. Even though Pat was quite a bit bigger, Stanley managed to lift him a few inches off the ground. Then with a gasp, he put him down.

"Okay," he wheezed, "let's see your brother lift him higher than that!"

Everyone stared at Dennis, who suddenly didn't look so confident.

"Hey, Einstein," he said meekly. "Are you sure about this?"

"No backing out!" Stanley cried.

"We're not backing out," Einstein said. "Dennis, I know you can do it. All you have to do is pick the right spot."

Can you solve the mystery? How can Dennis lift Pat higher than Stanley? Where is the right spot?

"What's the right spot?" Dennis asked.

"In the water, of course," Einstein said. "Pat, will you follow me and Dennis please?"

Then to the amazement of everyone, including Stanley, Einstein led Pat and Dennis into the lake until the water was over Pat's waist and up to Dennis's shoulders.

"Okay, Dennis," Einstein instructed. "Grab Pat around the legs and lift."

Dennis bent over until his head was under water. A second later, Pat rose up out of the lake and Dennis reappeared with his arms around Pat's legs.

A loud cheer came from the kids on the

beach. It was clear to everyone that Dennis had lifted Pat twice as high as Stanley had. The loudest cheer was coming from Paloma.

"Yay, Dennis!" she shouted.

"That's a trick!" Stanley cried angrily."Another Anderson trick!"

"It's no trick," Einstein said calmly as he, Pat, and Dennis waded back to the beach. "It's science. Pat floats because, like most people, he has a lower specific gravity than water," he explained. "So it was easy for Dennis to lift him. You could have done the same thing yourself."

"You . . .you . . .show-off!" Stanley spluttered, then he stalked away.

"That was cool," Dennis said, grinning broadly. "I'm super strong!"

"Yeah, Einstein, that was cool," Paloma agreed. "And not just because you were in the water."

"Hah!" Pat laughed, kind of loudly, at Paloma's joke. "That was funny."

"I guess," Paloma replied, giving Pat a puzzled look.

"Hey!" Einstein said. "That reminds me of something. Why were the commuters always wet?"

"Oh, no," Paloma replied. "Please, don't!"

But Einstein was already laughing as he said, "Because they were in a car pool!"

From: Einstein Anderson

To: Science Geeks

Experiment: What floats your boat?

If you've ever been swimming or even taken a bath, you know that an object (like you) in water seems to weigh less than the same object in air. That's how Dennis was able to lift Pat. But how does this work?

It's a principle that was discovered by a great scientist from Ancient Greece called Archimedes. He discovered that a body in water is acted upon by a force called "buoyancy," which exerts an upward force equal to the weight of the water displaced by the body. So, if an object weighs less than water, it floats. If it weighs more than water, it sinks. But hold on -- there's another factor here, not just weight. We know that steel is very heavy, heavier than water, but boats made of steel can float. Why is that?

Let's do an experiment with clay and find out.

Here's what you need:

- Modeling clay
- A bowl or tub of water
- 20-30 paper clips
- 20-30 pennies
- A friend

Divide your clay into two equal pieces. Shape one piece into a round ball and put it into the bowl of water. What happens? Think with your friend about what you could do that would make the same amount of clay float. Try out some of your ideas.

That's right! If you shape the clay like a bowl or a boat with high sides, it will float. Make a shape that floats and notice that some of your boat is under water. Mark the side of your clay boat to show the water line on the outside. Now add paper clips or pennies, one at a time, to your boat. What happens to your water line when you add pennies or paper clips? How many items can you put into your boat before it sinks? If you change the shape of your clay boat, will it hold more paper clips or pennies before it sinks?

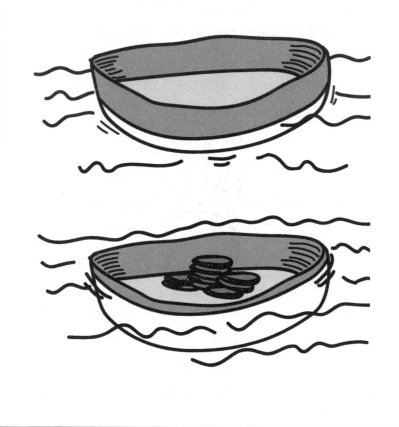

The Science Solution:

The shape matters. The reason why your clay bowl or boat could float is because the inside of the shape was full of air, which is less dense and weighs much less than clay. When you put your clay into the water, it "displaces," or pushes aside, a volume of water that weighs the same as the clay. A ball of clay sinks because all of its weight is concentrated in a small area and clay is more dense and heavier than water. But if you spread out the clay, it displaces the same amount of water, but over a larger area, and it's less dense because it contains both clay and air. The water line you drew on the side of the boat shows the amount of water the boat displaced. When you add paper clips or pennies, the boat sinks deeper to displace the weight of the boat plus those items. Eventually, if there are too many things in the boat, water comes in over the top and displaces all the air, and it will sink.

Try a contest with a friend: Using the same amount of clay, who can make a boat that will float more paper clips?

The story is that Archimedes was taking a bath when he figured out this principle. He was so excited by his discovery that he jumped out of the tub and ran into the street shouting "Eureka!" Which means, "I've found it!" in ancient Greek. Don't try that at home. Or at least put your clothes on, before you run outside!

The Red Bull

"Come on, Einstein!" Dennis yelled from his seat in the back of the pickup. He'd started teasing Einstein about his name, the way Paloma did. But Einstein didn't mind. He liked his nickname, no matter how anyone said it. He just closed the back door to the house and walked over to the truck.

It was an early Saturday morning, the first

week of August. Every so often, Matt Anderson liked to take his sons with him when he made calls on sick farm animals and this was one of those times.

Einstein enjoyed seeing his dad at work. Dr. Anderson had the reputation of being the best veterinarian in the county. He always seemed to know just what to do to calm down an animal during an examination. His hands were gentle, and he worked quickly to find out what was wrong.

"Hurry up!" Dennis exclaimed as Einstein climbed in the back. "I want to get there before the rooster goes cock-a-doodle-doo!"

"I think we're too late for that," their dad laughed from the driver's seat. "But roosters can crow any time, so you still might hear it."

Dr. Anderson put the pickup in gear and they pulled out of the driveway. He was taking them to the Joneses' farm to check on some sick cows.

"Hey, Einstein," Dennis said, looking over with a smile. "You have two different colored socks on. You must be color-blind!"

Einstein looked at his feet. Dennis was right. He had a blue sock on one foot and a green sock on the other.

"Can you see what color my shirt is?" Dennis asked jokingly, pointing to his bright red T-shirt.

"Yes, I can see it," Einstein told his little brother. "I'm not color-blind. I just wasn't looking when I put on my socks. But it's an interesting point," he continued.

Dennis rolled his eyes. "No, it's not!" he protested.

"How do you know? I haven't said anything yet." Einstein said.

"You always think it's interesting to talk about science," Dennis replied.

"Well, it is," Einstein said. "For instance, did you know that dogs can only see blue and yellow? A lot of animals see colors differently than we do."

"That's not that interesting," Dennis insisted.

"Okay, then how about this?" Einstein went on. "Since we're talking about colors, what's black and white and red all over?"

"Everyone knows that one!" Dennis cried. "It's a newspaper!"

"No," Einstein shook his head. "A zebra that fell in cranberry juice!"

"Dad!" Dennis yelled. "Turn on the radio! Please!"

The Joneses' farm wasn't far outside of town. When they pulled into the long dirt drive that

led up to the farmhouse, Mr. Jones was waiting—a big round-faced man wearing a plaid
shirt and blue jeans.

They got out of the van, and Dr. Anderson
and Carl Jones said their hellos and shook
hands. Then they turned toward the dairy barn.

"Are you coming, Dennis?" Matt Anderson
asked his sons.

"I want to see the rooster," Dennis replied
and ran off in the other direction.

"Well, I want to come," Einstein said.

Carl Jones led them to the dairy barn and Dr.
Anderson examined four cows that were being
kept separate from the rest of the herd. Einstein watched his father carefully, trying to learn
all he could about animals and medicine. Finally the exam was done.

"They seem to be a lot better, Carl," Dr. Anderson said. "Just keep giving them the pills I

gave you on Wednesday."

"That's a relief," Mr. Jones said, with a broad smile. "While you're here, could you take a look at Ajax?"

Ajax was Mr. Jones's prize bull who always won the blue ribbon at the county fair.

"What's wrong with Ajax?" Dr. Anderson asked.

"Nothing," Mr. Jones replied. "At least, he seems okay most of the time. But sometimes, when I go up to the pasture, he's all sweaty and irritable. And I think he's been charging the fence. If he keeps behaving this way, I won't be able to enter him in the county fair. Joe Burns would love that. Then Apollo would finally win."

"Joe Burns owns the farm over the hill," Dr. Anderson explained to Einstein. "His bull, Apollo, always comes in second to Ajax. It's the biggest rivalry in the county, like the Red Sox versus the Yankees. Well, let's take a look at Ajax and see what's wrong."

On one side of the Joneses' farm was a large grassy slope and near the top of the hill was a fenced-in pasture where they kept their prize bull, Ajax. Carl Jones led them up the dirt road to the pasture. As they climbed the slope they saw that Dennis was already up there, standing just outside of the fence. Ajax, a beautiful black and white bull, was standing inside the fence not far away.

"Hey!" Dennis shouted when he saw his father and Einstein. "This bull is crazy."

"What do you mean?" Dr. Anderson asked, looking concerned.

"I was just standing here—I swear!" Dennis cried, "And he started charging the fence for no reason."

"That's what I was saying," Carl Jones told them. "What do you think, Matt? Can you help him? Do you think he's sick?"

They all looked at Ajax. The bull seemed angry and nervous. But Dr. Anderson calmly opened the gate and walked into the pasture, talking in a soothing way until he was able to examine the animal. At the end of the examination he came back out and closed the gate behind him.

"Ajax is in fine shape," he told Mr. Jones. "But something is getting him worked up."

"Well, I wish I knew what it was," Carl Jones replied. "And you know, the fair is just a couple of weeks away. If he's acting like this, there's no way I can enter him."

Matt Anderson thought for a moment. "Carl, I wonder if you would mind letting Adam stay with the bull this morning to watch him. You

just go on about your chores the way you always do."

"That'd be fine," said Mr. Jones. "But will Adam know what to look for?"

"Don't worry, there's no one better at getting to the bottom of a science mystery," answered Dr. Anderson. "After all, they don't call him Einstein for nothing."

"Hey, what about me?" Dennis asked. "I can solve mysteries, too!"

Dr. Anderson laughed. "Well, if you promise to listen to Einstein, you can stay, too."

Dennis looked unhappy, but he nodded to his father.

"Okay, boys, I'll be back before lunch," Dr. Anderson told them.

"Don't worry about them," Carl Jones said. "They'll be just fine."

Their father and Carl Jones walked back

down the hill, leaving Einstein and Dennis standing by the pasture fence.

"Okay!" Dennis said, sounding excited. "What do we do now? How do we solve the mystery?"

"We watch and wait," Einstein said, looking around.

"Watch and wait?" Dennis protested. "That's boring!"

"A good scientist knows how to observe patiently," Einstein replied.

"Yuck!" Dennis answered.

"Come on," Einstein said, pointing to a small stand of trees not too far away. "Let's wait in the shade. Plus, I think we shouldn't be seen."

"Why not?" Dennis asked, but Einstein was already headed for the trees. Dennis followed him reluctantly. The two boys sat down in the long grass in the shade of the trees and watched

Ajax, who didn't seem to be doing very much except brushing off flies with his tail. In fact, he seemed just fine. Soon, in spite of what he'd said about the importance of observation, Einstein got very drowsy and lay down in the soft grass. In a few minutes, he was fast asleep.

He didn't know how long he'd been asleep when he was awakened by a loud shout from Dennis. Jumping up, he saw his little brother over by the pasture fence. Ajax was snorting and pawing the ground, ready to charge.

"Dennis, get away from the fence!" he shouted and ran toward his brother.

As he did, he saw another figure run up, as if from nowhere—a boy maybe a little older than he was, and a lot bigger. He was wearing a baseball cap, a plaid work shirt, and blue jeans.

"Yeah!" the boy shouted, as he also ran toward Dennis. "Get away from that bull! He's dangerous!"

Einstein reached Dennis just as Ajax put down his head and charged the fence. Luckily, the bull pulled up and stopped before he ran into the wooden railings.

"Are you okay?" Einstein asked Dennis, grabbing him by the shoulders.

"Yeah, sure," his brother replied.

"Yeah, are you okay?" the other boy asked as he ran up. Einstein noticed he was fumbling, as if he was putting something away in the pocket of his blue jeans. The boy was even taller up close and Einstein could see he was probably old enough to be in high school.

"Good thing I was walking by, or that crazy Ajax would have hurt you," the boy said. "I'm Jesse Burns. I live right over there." He pointed over the crest of the hill.

"I'm Adam," Einstein said. He didn't usually

use his nickname with strangers. "And this is my brother, Dennis. We're uh, we're visiting."

"Everyone calls him Einstein," Dennis added.

"Dennis, what happened?" Einstein asked him. "I . . . uh . . . I didn't see."

"You were asleep, you mean," Dennis replied. "I don't know what happened. All of a sudden Ajax got really wild, kind of stomping around, moving his head from side to side. I came over to the fence to get a better look. That's when he started to charge."

"I see," Einstein said. "And where were you, Jesse?"

"I was just walking by," the boy replied, getting red in the face. "I cut through here all the time. Mr. Jones doesn't mind."

"You said your name was Burns," Einstein asked him. "So your dad is Joe Burns, who owns the bull named Apollo."

"Yeah, so what?" Jesse said. "Apollo is a lot better than this old hunk of hamburger."

"So, I think you've been doing something to get Ajax angry," Einstein said evenly. "So Mr. Jones wouldn't enter him in the county fair this year and Apollo would win."

"You're nuts!" Jesse shouted. "I didn't make that bull charge. It was your brother's red shirt. That got him angry. He should know better than to wear red around a bull."

"It's not my fault!" Dennis said. Then he added, "It's not, is it, Einstein?"

"It certainly is not," Einstein told him. "I'm certain."

Can you solve the mystery? How could Einstein know that Dennis's red shirt didn't get Ajax angry?

"I know it's not your shirt," Einstein told Dennis, "because bulls are color-blind. All the stories you heard about bulls getting angry when they see red are just not true. Something else got Ajax angry—and I bet it's whatever Jesse has in his pocket."

"There's nothing in my pocket," Jesse said angrily, his face turning red. Then slowly, he added, "Just my cell phone."

The boy took out his phone and showed it to them. It was bright, polished chrome and glinted in the sun.

"Could I see it?" Einstein asked, holding out his hand.

"No!" Jesse said angrily. "You want to try

and take it?"

He held the phone high in the air, daring Einstein to try to get it. But as he did, the shiny case caught the sunlight and reflected it right at Ajax's head. The big bull snorted angrily and shook its head.

"Never mind," Einstein said. "You just proved my point. That phone case is like a mirror. You've been using it to reflect light into Ajax's eyes and get him angry," Einstein said calmly.

"So?" Jesse replied. "Apollo is still a better bull. I don't care what those stupid judges say."

Then he turned on his heel and strode off over the hill.

"Well, I guess we solved that mystery, right, Einstein?" Dennis said as the two of them walked down the hill toward the farmhouse.

"We sure did," Einstein replied. "But there's

one more riddle you need to solve."

Dennis knew what was coming. "Oh, no!" he groaned.

"Why couldn't the farmer tell the difference between a bull and a cow?" Einstein asked, already laughing. Then without waiting for Dennis to answer, he added, "Because, whichever one he picked, the cow was always the udder one!"

From: Einstein Anderson
To: Science Geeks
Experiment: Living Color

Have you ever seen a rainbow? Rainbows happen after a storm when sunlight hits raindrops at a particular angle. Sunlight looks white. But white light is actually made up of an array of different colors called the "visible spectrum." When a sunbeam hits raindrops at just the right angle, the different colors separate and we can see them. You might have seen this effect when light passes through a triangular piece of glass called a prism. Each color is a different wavelength and raindrops and prisms separate the spectrum of light into all its wavelengths.

So if raindrops can break white light into a rainbow of colors, can we put colors together to make different colors, or even white? Let's do an experiment with a color wheel and find out.

Here's what you need:

- A white paper plate or a circle cut out of white poster board
- A ruler
- A pencil
- Paint or colored pencils
- A piece of string about 18" long
- A pair of scissors

Using your ruler and pencil, divide the circle into 12 sections that are as close to equal as you can make them.

First, color in the three primary colors, red, yellow and blue. One primary color should be on the top of the circle at 12 o'clock, one near the bottom at 4 o'clock, and the third about the same distance from the bottom, but on the other side, at 8 o'clock. Next, mix each pair of the primary colors (red and blue, blue and yellow, yellow and red) to form the three secondary colors, purple, green and orange. Put these in-between the two colors they are made of.

Finally, make the tertiary colors by blending a primary with the adjacent secondary. See how seamlessly you can blend the colors.

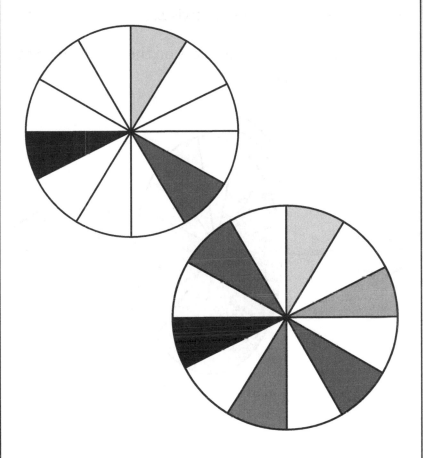

When your color wheel is finished, poke a small hole in the center and thread the string halfway through.

Now comes the fun part! Holding both ends of the string, swing your color wheel in a circle so the string gets twisted. Then grip each end of the string and pull – the wheel will spin as it un-twists.

What do you see? Do the colors on the wheel blend to make another color?

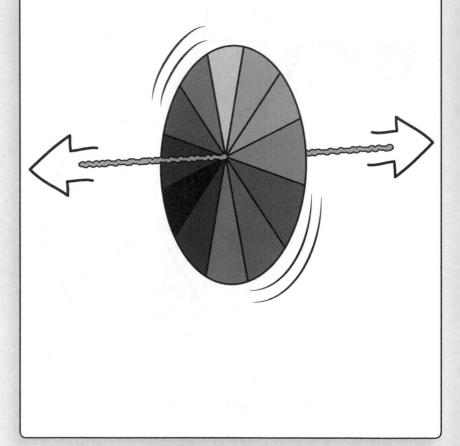

The Science Solution:

As the color wheel spins, your brain blends the moving colors together. If you made a color wheel with just two primary colors in alternating sections, it would make the secondary color when you spin it — for example, if you used yellow and blue, the spinning wheel would look green. If you have time, try some different color combinations and see what you get!

And what about Ajax the bull -- is he color blind? The answer is no, not for a bull. Different animals see colors differently. As humans, most of us can see all the colors in the rainbow, but no more. That's why we call these wavelengths of light the "visible spectrum." Some animals can see colors we can't see, and some see only a few of the colors in our visible spectrum. For example, Dr. Gerald Jacobs at the University of California has shown that dogs can see blue and yellow and tell the difference between them, but they can't see any difference between red and green. And bees use their ultra-sensitive color vision to follow ultra-violet markings we can't see, and find their way deep into flowers.

Lightning Never Lies

It was only one week before school would start again. Soccer camp was over for Einstein and day camp was over for Dennis. There was just one week left to get in more swimming, ball playing, and summer fun. One week left with no homework, no studying, no tests, and no report cards.

To celebrate the last week of freedom, Einstein's family decided to have a picnic at Carter Lake. Einstein went up the night before to stay with Paloma at her Aunt Camilla's farmhouse.

"Well, Einstein," Paloma said, after he arrived. "What great science project do you have in mind?"

Einstein smiled and simply said, "Cookies."

"Cookies?" Paloma laughed. "That's not science, Einstein, that's baking."

"Baking is science, Paloma," Einstein replied. "Here, take a look."

He held up his smart phone, which was open to a page about the science of baking.

"Baking is all about chemistry," he explained. "Even if bakers don't always know it. For example, the protein in flour is what gives bread or cake its form. Different types of flour have different amounts of protein. Cake flour makes a lighter loaf than bread flour."

Paloma looked impressed. "Okay, I never knew that," she said. "My aunt has a great chocolate chip cookie recipe. She's a scientist and a baker. Maybe we can have her write the recipe as a chemical formula for the batter."

"Hey, that reminds me," Einstein said with a grin. "What do baseball and cookies have in common?"

"That's an easy one, Einstein," Paloma laughed, in spite of herself. "They both have batters at the plate."

Paloma's aunt Camilla had put her cookie recipe on Pinterest, so she set up her laptop in the kitchen so they could read it. Then she helped get the two of them set up with all the ingredients they needed. Soon they were mixing a large bowl of cookie batter. By the time they were done, they had baked over four dozen cookies. Paloma put a tweet about it with a photo on her Twitter feed.

The next day dawned cloudy and it looked like rain. But Einstein's mom called to say they were coming up anyway and bringing Einstein's friend Pat. Paloma's parents were coming, and with Aunt Camilla, they had a nice big group. Everyone arrived early and they went out to the lake to set up on a couple of big picnic tables. The kids were already wearing their bathing suits, hoping to get in a swim before lunch.

The picnic tables were loaded with fried chicken, cold cuts, salads, watermelon, and delicious bread. There was even a peanut butter and jelly sandwich for Einstein's brother Dennis, who ate nothing else. Einstein and Paloma proudly set out two plates of their cookies for dessert.

"Nice cookies, Einstein," someone sneered as Einstein carefully set down the last plate. It was Stanley Roberts.

"Stanley!" Paloma sighed. "Who invited you?"

"Einstein's parents," Stanley replied with a grin. "They invited my whole family. I guess they think I'll be a good influence on you."

"Good influence?" Paloma sputtered. "I . . . I . . . That's nuts!"

"Come on, Paloma," Einstein said gently. He didn't like Stanley much but sometimes he felt sorry for him. "Stanley's our guest, I guess."

"Gee, thanks, Einstein," Stanley said in a mean way. "That's really big of you. As a reward, you can have one of my super scientific cookies. They're a lot better than those hockey pucks you and Paloma baked."

He pointed to a plate of rather burnt-looking cookies at the other end of the table.

"That's just like you, Stanley," Paloma said

angrily. "You always have to prove you're better than me or Einstein. You saw my Twitter post and so you had to go and make cookies of your own."

"There's no law against that, is there?" Stanley asked.

"No, there's no law," Paloma replied. "But there is a law against poisoning people and I'm sure that's what your cookies do."

"My super scientific cookies," Stanley corrected her.

"What exactly makes your cookies so scientific?" Einstein asked, sounding curious.

"Hah!" Stanley replied. "Trying to trick me into revealing my secret cookie recipe? It's going to make me a billionaire."

"Come on, Einstein," Paloma said. "Let's go swimming. And don't eat any of Stanley's cookies or you'll sink like a stone."

"You'll see!" Stanley said as Paloma and Einstein walked away. "Before the day is out you'll be begging me for a cookie."

They ignored him and ran to the beach, where they found Pat and Dennis already in the water. Soon they were swimming and splashing and they forgot about Stanley altogether. In a little while Aunt Camilla came down to call them to lunch. The sky had gotten darker while they were in the water.

"Come on, kids!" she called. "Let's eat before it rains."

The three families sat down at two long tables and everyone dug in. Stanley sat at a separate table with his family and didn't even say a word to Einstein and Paloma. They concentrated on the food and didn't stop eating until they were stuffed. They were both about to reach for one of their cookies, when Einstein's mom looked up at the darkening skies.

"I think those are thunder clouds," she said in a loud voice. "We better get to shelter."

Einstein looked up. His mom was right. Those were tall, very dark cumulonimbus clouds and they were coming over the lake, fast. Suddenly, there was a loud noise like distant thunder. Paloma checked the weather app on her phone. "There's an alert for dangerous storms in this area," she announced.

"My house is just up the hill," Aunt Camilla announced. "Come on."

Everyone grabbed whatever they could and headed up the hill to Aunt Camilla's.

"The cookies!" Paloma shouted as her aunt took her by the arm.

"I put some plastic over them," Camilla replied. "I'd rather lose some cookies than have you be struck by lightning."

75

In less than a minute, everyone was safely in Aunt Camilla's house. Einstein noticed that Stanley was the last one in the front door. The grown-ups tried to reassemble as much of the lunch as they could on her large kitchen table. Pat and Dennis watched the storm from the windows in the living room. Meanwhile, Einstein and Paloma tracked the radar image of the storm on her aunt's computer.

The dark clouds came over quickly and the sky was almost black. In the middle there was a single flash of lightning. Paloma started counting slowly, "One Mississippi, two Mississippi, three Miss . . ."

Einstein knew she was counting to see how far away the lightning had been. When she reached five Mississippis a loud rumble of thunder shook the house.

"Hmm, a mile away," he said and Paloma

nodded. They knew that light traveled much faster than sound. Light traveled at 186,000 miles per second. The light from a lightning strike miles away would reach your eye in a tiny fraction of a second. But sound traveled much more slowly. It took five seconds for sound to travel one mile. So if you saw lightning and counted the seconds until you heard the thunder, you could tell how long the sound had traveled, and how far away it was.

Luckily that one lightning strike was the only one from the storm as it passed overhead. Like thunder storms often do, this one moved away as quickly as it came. Soon the sky was clear and blue.

"Come on, everyone," Aunt Camilla said. "Let's go rescue our picnic."

They all went back down the hill to the picnic tables. Paloma rushed to the spot where the cookies had been and pulled back the plastic tablecloth. They were gone! It wasn't just the cookies, but the plates they were on were missing, too.

"Oh no!" Paloma cried. "What happened to them, Einstein?"

Einstein didn't say anything, just looked at the picnic table and scratched his chin thoughtfully.

"Gee, that's too bad," Stanley said, walking up behind them. "I guess everyone will have to eat my super scientific cookies now."

"The wind didn't do it," Paloma declared, looking around. "If it were the wind, there'd be cookies scattered all over the place and we'd see the plates or at least pieces of them."

She turned and fixed Stanley with an angry glare.

"It was you!" she said, pointing her finger.

"You stole our cookies."

"That's nuts!" Stanley shot back. "Why would I want your stupid cookies, when I have my super scientific cookies? Besides, I saw what happened to your precious cookies—a bear ate them."

"A bear?" Paloma and Einstein said at the same time.

Einstein's mom heard the fuss and walked over.

"What's going on here?" she asked. Paloma quickly told her about the missing cookies, blaming Stanley.

"Oh, I'm sure Stanley didn't take the cookies," Emily Anderson said gently. "On the other hand, there aren't a lot of bears around here. Are you sure that's what you saw, Stanley?"

"Yes, I'm sure, Mrs. Anderson," Stanley insisted, with a look of pure innocence on his face. "I was watching from up at the house, through the front door. I remember because there was a loud clap of thunder and I looked down at the lake. Then there was a flash of lightning, and in the flash I saw a big black bear stealing the cookies."

"And did the bear steal the plates, too?" Paloma asked sarcastically.

"He must have," Stanley replied, looking innocent. "There's no other explanation."

"Paloma," Mrs. Anderson said, "I know it doesn't make sense, but if that's what Stanley says, it must be true."

"Einstein!" Paloma protested. "Aren't you going to say anything?"

Einstein looked thoughtfully at the table again. Then he turned to Stanley. "Stanley, are you sure that's exactly how it happened? You heard the thunder and then saw the bear by the flash of lightning?"

"Absolutely," Stanley insisted. "That's exactly how it happened."

"Well, that's too bad," Einstein replied, shaking his head. "I didn't see you take the cookies, Stanley, but I know you're not telling the truth about seeing the bear take them."

Can you solve the mystery? How does Einstein know Stanley is not telling the truth?

"How do you know that?" Stanley demanded.

"The way you told the story just isn't possible," Einstein said. "You see, a flash from a bolt of lightning travels at the speed of light, 186,000 miles per second."

Paloma had her smart phone out and quickly displayed that exact information on the screen.

"Yeah, so what?" Stanley said, reading what it said. "The speed of light, everybody knows that."

"So, sound travels about six hundred miles per second—a lot slower," answered Einstein.

Stanley just looked at them dumbly, but Mrs. Anderson couldn't help smiling.

Paloma jumped in. "Don't you see?" she asked impatiently. "You said you heard the thunder and then you saw the lightning. But it couldn't have happened in that order. First you see the lightning, then you hear the thunder."

"And there was only one thunderclap," Einstein added. "Everyone noticed that."

"Stanley," Einstein's mom said, gently but firmly. "Do you think you might have been . . . mistaken about the bear?"

"Uh. . . maybe," Stanley said slowly. "Uh . . . yeah, in fact, now I remember, Mrs. Anderson! I put the cookies in the shed over there," he said, like it was the nicest thing in the world he could have done. "I just wanted to protect them!"

"Well, dear, that was very thoughtful of you," Mrs. Anderson said with a smile. "But now, why don't you go get them and put them back on the table? It's time for dessert."

"Good work, Einstein," Paloma said, as they watched Stanley trudge over to the shed to get the cookies.

"Thanks," Einstein said. "But there's one more thing I wanted to say to Stanley."

"What's that?" Paloma asked, without thinking.

"He shouldn't feel bad. That's just the way the cookie crumbles!"Einstein said with a laugh.

From: Einstein Anderson
To: Science Geeks
Experiment: A Tempest in a Teapot – or a Cloud in a Bottle

Did you ever wonder what clouds are made of and why they form? Let's do an experiment where we make a cloud in a bottle.

Warning: You need an adult to help you with this experiment because it requires lighting matches.

Here's what you need:

- A 2-liter plastic soda bottle
- Warm water
- Matches

Fill the bottle 1/3 full with warm water. Put on the cap and squeeze the bottle. What happens? Next, take off the cap and have an adult light a match and hold it near the mouth of the bottle. Drop the match into the bottle and quickly screw on the cap. Now, squeeze the bottle again and let it go. What happens this time? Why do you think you saw a cloud this time? Why did it go away?

The Science Solution:

It takes three things to make a visible cloud: Water vapor, particles for the vapor to cling to, and a drop in air pressure.

When you squeeze the plastic bottle, it raises the air pressure (the same amount of air is pressed into a smaller area) and when you release the bottle, the pressure drops. Increasing pressure raises the temperature of water, so when you squeezed the bottle, the warmer air was able to hold more water vapor. But you can't see the cloud because there are no particles for the vapor to cling to.

When you add smoke from the match, the water vapor clings to the smoke particles. When you stop squeezing the bottle, it drops the air pressure and a cloud forms. You've got all three conditions that are necessary to make a cloud in your bottle:

1. Water vapor from the warm water,
2. Smoke particles for the water to cling to, and
3. A drop in air pressure.

Who says you can't control the weather?

The Baseball Angle

The afternoon after the big thunder-storm was clear and warm. Einstein and Paloma quickly forgot about Stanley's attempt to steal the cookies. They went swimming again with Pat and Dennis and a bunch of the other kids, including Jamal Henry, Carla Korzak, and Tom Patel. Later they went up to the baseball field to play softball. They didn't have enough kids for two full teams, so they just took turns batting and fielding. Pat was the star pitcher for the town team, so of course he took the pitcher's mound.

Paloma was first at bat.

"C'mon, Pat," she dared him, crouching in her stance by home plate. "Let's see how good you really are."

Pat was used to pitching overhand with a hardball, fast and straight over the plate. But as usual, he acted a little strange whenever he was around Paloma. Now he had to pitch to her. She stood over the plate with a look of fierce concentration, waving her bat over her shoulder. She scuffed the dirt in the batter's box with the soles of her red canvas sneakers. Meanwhile, Pat shifted nervously on the pitcher's mound.

"Come on!" Paloma repeated, impatient as

usual. "What are you scared of?"

Pat didn't say anything. He just took a deep breath and carefully threw a nice, easy underhand softball pitch that hung over home plate. Paloma took a mighty swing and sent the ball into right field, where Dennis went running after it.

Dennis was only eight but he ran quickly and easily and scooped the line drive up in his glove. Then he grabbed the ball, cocked his arm, and threw it toward Jamal, who was playing first base. The ball fell far short and Jamal had to run off the base to get it. Paloma was easily safe.

"Come on, Pat," she called to the pitcher's mound. "Next time, don't give me such easy pitches."

Pat look flustered. "Uh, okay, I was just . . ." his voice trailed off and he looked away. Paloma didn't seem to notice.

Meanwhile, Einstein, who was playing center field, jogged over to his brother.

"You have to throw it in a higher curve," he told him, "if you want to reach the infield."

"But in the major leagues the pitcher throws the ball fast and straight," Dennis replied.

"No, it only looks straight," Einstein explained. "Because it's traveling so fast. Every thrown ball follows a curve of some kind, to go against the pull of gravity. The faster and harder you throw, the flatter the curve can be."

Dennis just nodded. As usual, he didn't understand much of what Einstein said, but he knew his big brother was probably right.

They all took turns batting, and then Pat tossed the ball to Einstein, who was still in the outfield. He threw the ball in a high curve so it fell right into Einstein's glove.

"Here, Einstein, you pitch to me," he said.

Einstein caught the ball and walked over to the pitcher's mound. He knew he was nothing like the athlete Pat was, but he was ready to give it a try. Of course, Stanley chose that moment to show up at the field. He was wearing shorts, a white T-shirt, and a blue baseball cap with the words STANTASTIC INDUSTRIES on it.

"Look who's pitching!" he called from behind the backstop. "Let's see how you use science now, Einstein!"

"Come on, Einstein," Pat said, holding the bat over his shoulder. "Never mind him. Just nice and easy, over the plate."

"Yeah, Einstein," Paloma cried from her spot at first base. "He's just jealous, as usual."

"Me? Jealous of him?" Stanley sneered. "I just want to see the genius at work."

Einstein set himself, then did just as Pat suggested, throwing the ball underhand so it curved gently up then fell right across home plate. Pat swung and with a mighty crack, he sent the ball sailing far over the outfield.

"Hah!" Stanley laughed from behind home plate. "Some pitcher you are."

"It's just batting practice," Paloma shouted angrily.

"Practice, shmactice," Stanley replied. "It's just like I said. Einstein always claims he can use science to solve anything, but here's one place he's wrong."

"He's not wrong," Paloma replied angrily. "Right, Einstein?"

Einstein shrugged. He didn't want to get in another contest with Stanley again—not twice in the same day.

"Look, let's just play ball," he said, trying to sound as friendly as he could.

"Oh, sure," Stanley said with a laugh. He walked out to the pitcher's mound with a swagger. "I guess you're not really a genius, are you?"

"I never said I was a genius," Einstein replied.

"Oh, yeah, you just call yourself Einstein, that's all," Stanley sneered. He was just a few feet from the mound. Everyone else had gathered around—Paloma, Pat, Dennis, Carla, Tom, Jamal, and the rest of the kids. Dennis gave him a worried look.

"Come on, uh, Adam," he said. "Show him! Show him, you know . . . science stuff!"

Now Einstein knew that this was serious. Dennis never called him by his real name. He felt the eyes of all the kids were on him.

"Dennis, I don't want to show Stanley, or anyone else, science stuff," he explained, picking his words carefully. "I don't use science to show off. I just use it because it's fun."

"Yeah, right!" Stanley said. "He does it 'cause it's fun. How lame can you get? Everyone knows you're always showing off, Einstein!"

Einstein heard Carla and Jamal and the other kids giggle a little. Dennis wouldn't meet his eyes and looked at the ground. Paloma gave him a pleading look.

"Okay," he said with a sigh. "If it will shut you up, Stanley, I'll show you how I can use science to beat you in sports."

"You will?" Stanley said, sounding surprised. Then he quickly added, "I mean, no, you won't!"

"Yes, I will," Einstein told him. "But then you have to promise to leave me alone, at least for the rest of the summer. Is it a deal?"

"Sure, no problem," Stanley said. "But what if you don't? What do I get?"

Einstein thought a moment. "If I don't beat

you, Stanley, then everyone agrees to call you Einstein instead of me!"

"No!" Paloma gasped.

"Einstein Roberts," Stanley said gleefully. "I like the sound of that!"

"You won't win!" Dennis declared. Then he added in a worried tone, "Will he, Einstein?"

"You can use science anyplace," Einstein said, without answering him, "even in sports." He pushed his glasses back on his nose. "And so, Stanley, I challenge you to a throwing contest."

"A throwing contest?" Stanley asked, sounding confused. "What's scientific about that?"

"You'll see," Einstein told him, with a smile. "And I'll give you a big advantage. I'll let Pat throw for you."

"What?" Stanley and everyone else started talking at once.

"Me?" Pat's mouth dropped open in surprise. "Why me?"

"Because I don't want Stanley to say I cheated, that's why," Einstein explained. "But there's one condition."

"I knew it!" Stanley shouted. "Here comes the trick!"

"It's no trick," Einstein told him. "You said I could use science. So here's the scientific contest. We each throw the ball as far as we can, but we have to throw differently."

"And how are you going to throw it, Einstein?" Stanley challenged him. "With a cannon?"

"Oh, no," Einstein said. "I'm just going to use my arm and nothing else. I'll throw the ball from the same spot Pat does. We'll each get one chance, and I'll bet my throw goes farther."

Can you solve the mystery? What does Einstein know about throwing a ball that will help him throw farther than Pat?

"Here's the only rule," Einstein said. "Pat has to pitch the ball like he would in a regular game. I can throw any way I want."

"What kind of crazy rule is that?" Stanley asked, looking confused.

"It's a scientific rule," replied Einstein with a smile. "You said I could use science."

"Okay," Stanley grumbled. "But Pat, you'd better throw hard."

"Don't worry," Pat told him. "I don't want to be beat by anyone."

"I'll go first," Einstein said. Without hesitating, he took the ball, hauled back, and threw it in a nice, high curve. It landed by some trees at the very edge of the field.

"Hey, no fair losing the ball!" Stanley protested. But Jamal had already run after it. A few seconds later, he held it up.

"Here it is!" he shouted. He threw it back and Tom caught it, then handed it to Pat.

"Go ahead, Pat, show him who's the best baseball player in town," Stanley said, trying to sound confident.

Pat took his pitching stance, then let go with a mighty pitch. The ball left his hand like it really was shot out of a cannon and it rocketed toward the outfield. But it fell far short of Einstein's throw.

"It's a trick!" Stanley shouted. "Pat didn't try."

"Oh, I tried," Pat said rubbing his arm. "But everyone knows you can't throw to the outfield the same way you pitch to home plate."

"What do you mean everyone knows?" Stanley said.

"Well, every baseball player knows that," Pat replied.

"It's science," Einstein said. "It's like I was explaining to Dennis before. Gravity will pull down any ball you throw or hit. If you throw the ball straight out in a flat line the way Pat did, gravity will pull it down to the ground before it goes very far."

"But if you throw the ball straight up," Dennis said, "it will come down close to you."

"That's right," Einstein said. "Ball players learn the right angle for the right throw. It depends on their arm strength and how far the ball has to travel."

"But no one figures out the angle before they throw," Pat objected.

"No," Paloma told him, nodding excitedly. "You just know from experience."

"I do?" Pat beamed happily at Paloma's

words.

She gave him a puzzled look. "Uh, yeah," she told him. "'Cause, you know, you're good at sports."

"I am?" Pat replied. Now he looked very embarrassed and suddenly looked at his feet.

Paloma looked to Einstein, but he was still facing off with Stanley.

"You cheated!" Stanley shouted at Einstein. "I know you did."

"No, he didn't," Jamal told him. "He beat you fair and square just like he said he would—with science!"

"Come on, Stanley," Einstein said as nicely as he could. "You know, we don't have to compete all the time. There's plenty of science for everyone."

"Well, maybe," Stanley grumbled. He looked miserable, but then he shrugged. "Okay, I admit it. You beat me. I guess you're still Einstein, Einstein."

"That's right!" Dennis shouted gleefully. "And don't you forget it!"

From: Einstein Anderson
To: Science Geeks
**Experiment: Arc of Triumph –
Pitching to defeat gravity**

Hey, Science Geeks, here's an example of how science can help you become a winner! We all know that great athletes use both their bodies AND their brains to win at sports. So let's do an experiment that will give you an edge in any sport that involves throwing a ball.

Here's What you Need:

- An outdoor space or gym where you can throw far
- Two or three balls
- Four targets -- stakes or cones or items to mark distance

This experiment is more fun with friends, but you can do it by yourself as well.

First, set up your space. Mark a distance that is as far away as you could imagine throwing your ball, one that's about half that far, one that's about a quarter of the way to your far-away target, and one that is only about five feet from where you stand. Your goal is to throw the ball so that it gets as close as possible to each of the targets.

Throw your first ball and try to get it near the farthest target. If you fall short, try again, but this time, try throwing the ball higher, not just harder. Take turns with your friends, throwing two or three times to each target. Keep score by writing down who gets closest to the target on each throw. The winner is the person who gets closest to the most targets.

When you are not throwing, watch how your friends throw. What do you notice when they are trying to get the ball to go far? Do they throw differently when the target is closer? How? Do you throw differently to reach a near target?

The Science Solution:

Sir Isaac Newton, who lived in England in the late 1600's and early 1700's, is considered one of the greatest scientists of all time. He worked out three laws of motion that have been fundamental to the science of physics ever since. And, believe it or not, throwing a ball at a target demonstrates all three.

Have a look at this picture:

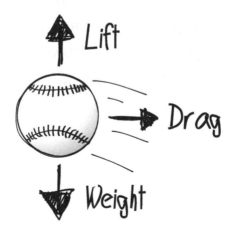

Forces on a Baseball

Lift

Drag

Weight

Newton's First Law of motion says that an object that is moving will continue to move at the same speed until it is slowed down by some resistance. The second law is a mathematical formula that helps you figure out the relationship between the weight or mass of the object and the amount of force that will make it move. The third law says that forces occur in pairs.

So, when you throw a ball, you exert a force that starts it moving. It would continue moving forever except for the force of gravity (or weight), which pulls the ball down toward the ground. The air the ball is flying through also exerts a force called "drag" that slows the ball. But, there is also the force called "lift," which opposes gravity and helps the ball rise. As a pitcher, the farther you want your ball to go, the more you want to encourage lift and resist the gravity that is pulling it toward the ground. If you want to hit your far target, you need to throw the ball up as well as out toward the target, and lift will help it keep going. If you want the ball to fall to the ground closer to you, you throw it with less force, but you also don't throw so high. That way gravity helps your ball drop to the nearby goal.

If you watch great ball players in action, you will notice that they adjust their throw to make the most of lift and gravity and they don't seem to think about it. But now you know how it works!

CPSIA information can be obtained at www.ICGtesting.com
Printed in the USA
LVOW01s1458071113

360407LV00012B/606/P